EXTRAORDINARY

The Genocide

KYLE FRANCIS

ISBN: 1515286150
ISBN 13: 9781515286158
Library of Congress Control Number: 2015912692
Createspace Independent Publishing Platform
North Charleston, South Carolina

CHAPTER 1

High school is a gamble. You either win big or lose terribly, but you might as well try. That's what freshman Aaron Anderson repeatedly whispered to himself on the bus.

A new school, a new day, a new selection of opportunities. Aaron was one who had trouble remaining positive, but as he did for every challenge he faced, he tried to keep a smile on his face.

The bus driver seems OK, he thought to himself. *Not too many of these kids look like they are felons. This should be an ordinary four years, of hell.*

He covered his face with his hands, hoping that when he looked again, he would appear at home, lying in bed. He peeked to see if his fantasy had come to life, but all he saw was a huge building painted a dark brown that almost matched his hair, surrounded by pointed steel fences outside the foggy window of the school bus. His new house was close to here, but it did not look this run down. His mother had warned

1

him that this would probably be a dangerous part of town, but moving here would be the best thing for him. She had said to him that they needed a fresh start.

The bus made a squeaky halt, and the rusty bus doors struggled to open. "A'ight, ev'ry bahdy. Heeya's ya stap!" snapped the bus driver with a cracked voice and a strong city accent. Aaron got up from his seat and saw a massive line of his fellow highschoolers trampling over each other to get out of the bus. The bell was ringing, which made the kids move faster, some even climbing on the ripped, leather seats, trying to reach the bus doors. Aaron was patient, but the line seemed endless. The walkway was empty about two minutes after the bell had rung. He rushed out the door and thanked the bus driver. The bus driver replied with a grunt.

He glanced at his directions, rushed to his class, and found the teacher slapping a ruler on her desk as he walked through the door. She was African American, and her hair was wrapped in a bun. A nameplate on her desk stated "Miss Birch."

"You're late!" she snapped. "Sit down."

Aaron sat at the only empty desk, which was in the back of the classroom in between a girl with short pink hair and facial piercings in her lip, nose, and both ears (who was giving him a look of death) and a boy with long black hair that covered some of his face (who was twirling a cigarette in his fingers under his desk). He turned and stared at Aaron with dark, purple-brown eyes. Aaron jumped a bit and quickly turned his head. It seemed inhuman to have such sinister-looking eyes. In the front sat an innocent-looking boy with blond hair and bright-green eyes which were revealed by him talking to the rest of the class about his answer. A few desks from the blond boy was a girl with long dark hair who looked to the back of the room often, taking a quick glance at the clock on the wall and immediately turning back. The rest were just a mix of jerky, drooling monsters who looked like they were about to fall asleep. Some of them looked so terrifying; they looked like they were recovering from a hangover. The thought made Aaron shiver.

"You scared, freshman?" The low, quiet whisper came from the boy next to Aaron, who was still twirling the cigarette in between his fingers.

"Hey, you're a freshman, too!" whispered Aaron.

The boy gave a quiet chuckle. "I am in ninth grade, yes, but you're scared. In this big school, you can't find a crevice to hide in. That's what a freshman is. A freshman is a coward." The boy's voice made him shiver again.

"Are you serious? I think you may need a dictionary." asked Aaron.

"Mr. Tempest and Mr. Anderson!" shouted Miss Birch. "Quit the chatter, or you will both serve detention! And throw that cigarette away!" The boy got up and threw the cigarette away. He returned and immediately opened his backpack pocket. Inside were about ten more cigarettes. He picked one up and started twirling it again. *I'm screwed*, Aaron thought.

It was lunchtime, and the cafeteria was a madhouse. People were shouting, fighting, and doing everything but eating. Aaron took his lunch tray and tried to find a good table. All the tables seemed to have their own groups. Aaron could not find any open spots except a table with four of the freshmen he had spotted in his first class. This included the cigarette boy, the pink-haired girl, the blond boy, and the clock girl. He walked to the table and sat down. They all stared at him. He tried not to stare back, but just simply smiled.

Aaron tried to start a conversation. "This place is like an asylum; am I right?" Aaron said. They just continued to stare at him, which made Aaron feel even more awkward.

"My God, finally someone said it," said the teacher's pet. "Once you enter those spiked gates, it seems like there is no return, but I see a light in you. I'm Ethan and these are my friends." He stood up from the table. "Ethan Winston." He reached out his hand toward Aaron. Aaron slowly shook it. Ethan sat back down, which was followed with more silence.

"Crystal," murmured the pink-haired girl. "Crystal Hardy."

Aaron nodded his head. "Good. Introduction…is always the first step to friendship. I guess this one came quicker than I expected."

The clock girl stood up, too. Now, Aaron could see her blue irises and small nose. "Jade. I'm Ethan's twin sister." She stuck her hand out toward Aaron. As Aaron shook it, he blushed a bit. They both sat back down.

The cigarette boy simply took out a cigarette from the same backpack pocket. He flicked the tip, and it lit. Aaron was amazed, at both the trick and how someone could be so addicted to cigarettes that he had to keep one in his hand at all times. The boy put the lit cigarette in his lips. "Like magic tricks, huh?" After a brief silence, he revealed his name. "Josh," he mumbled. He took the cigarette out of his mouth and said, "Joshua Tempest." He put the cigarette back in his mouth again and proceeded to smoke. It was odd of him to do this since smoking on campus was not allowed. He seemed like he didn't care.

"OK," Aaron murmured awkwardly. The bell rang, signaling the end of lunch. The five freshmen dumped their trash in the graffiti-covered trash can and walked to their next classes. Aaron walked behind the other four and noticed something near Ethan's feet. What he saw shocked him. He had wondered why Ethan was walking strange; now he spotted two prosthetic feet, which seemed as clear as day.

CHAPTER 2

School had finished at last. The day seemed to last a whole year. Aaron walked home from the bus stop and glanced at his new home. It seemed like a normal house, as if inside there was a neighbor whom he never met, but foreign enough to seem so different. He stepped through the door. His white Chihuahua, Snowball, ran toward him and jumped around. Aaron threw his backpack toward the wall and bent down to pet the dog. He didn't see his mom, though. He put Snowball on his shoulder and went to the counter. He saw a note with a small stain in the corner. It read:

Aaron,
I took a quick trip to the grocery store.
Don't make other plans, please!
I want to know how your first day went!
~Mom

Aaron smiled. He walked over to the leather couch, took some of the moving boxes off, and sat down. He picked up the remote and pointed it at the TV. It immediately turned to the news station. A news reporter announced, "Extraordinary beings are out there! They are real and dangerous! You could be attacked!" The program ended with the big bold letters "EPA" on the screen. Aaron didn't really know about the fuss with extraordinary beings. Supposedly they had dangerous, extraordinary powers—thus, their name—and they wanted to destroy humanity.

Aaron heard a knock on the door, followed by another two seconds after. Aaron opened the door. He saw his mother, her arms filled with groceries. "Hey! How was your day?" she said. She seemed like she didn't care about having her arms full and just wanted to hear about his day, but Aaron helped her anyway.

"Good, I guess," answered Aaron.

His mom gently set down the groceries on the floor. "I know," she said. "The first day is always the toughest." She adjusted her position to be more upright. "But you always get through it!" She hugged Aaron and put her hand on his face. "Have I ever told you that you look so much like your—" The doorbell rang.

Aaron's mom scooted to the door. She found two serious-faced men in navy-blue suits. A badge on their suits read "EPA." One was a hairless, bulky man whose hand rested beside his belt, which had a gun on it; the other had a buzz cut and a transparent electronic tablet with a silver pen. Aaron could see the screen from the other side. There were many different house addresses, ones with marks next to them and others without. The top of the screen showed the Andersons' address in big bold letters, without a mark. "Good afternoon, Mrs. Anderson," the one with the tablet said.

"Please," said Aaron's mom. "Call me Deborah."

"I'm Marcus. I'm a commander in the Extraordinary Protection Agency. Next to me is Griff." He pointed to the man beside him. "We have seen some…incidents with extraordinaries. Have you seen anything suspicious lately, ma'am?"

"Oh no; sorry, sir," answered Deborah. "I—" She paused for a moment. "I haven't seen anything."

"Well, that's good," said Marcus. He took his pen and tapped his tablet. Their address now had a mark beside it. "We'll be sure to keep you safe," he said. They walked from the door.

Deborah closed the door. She had a tear coming down from her eye. "Mom, are you all right?" Aaron asked.

"Yes," whispered Deborah, and then she stood up firmly. "I'm going to bed." She abandoned the groceries and walked to her room.

"Mom, it's five o'clock," said Aaron. His mom ignored him. Aaron sensed distress. Snowball was next to the window, growling. Aaron came over and looked out of the window. He saw Griff installing something in the wall of the house. It looked like a camera. Griff turned his head toward the window. Aaron ran into his bedroom. He lay on his bed and sighed. He smacked his head on his pillow. The dog jumped on the bed and began to lick Aaron's face. "Not now, Snowball," said Aaron. "Maybe it is a good idea to go to bed early today." Without any preparation, he fell asleep.

CHAPTER 3

Aaron was exhausted at school the next day, even though he had gone to bed early. His first class seemed to be emptier than it was the first day. He noticed that Josh was absent; he also noticed that Jade and Ethan weren't happy like they were yesterday. Jade seemed down, and Ethan looked almost irritated. Aaron wasn't happy about this. He wanted to leave. The clunky lunch bell started to ring.

At lunch, Aaron went to the same table he went to the previous day and sat down. Only Jade and Crystal were there, and both of them had their arms crossed in front of them on the table, heads resting on their arms.

"Where're Josh and Ethan?" asked Aaron.

Jade replied by burying her face in her arms.

"Josh is here today. He's just in detention," answered Crystal, looking up. "Ethan—well, he may be busy."

"Busy?" asked Aaron.

"Just...busy." answered Crystal. She put her head back down, and right at that moment, the bell rang. Aaron got up from the table and headed to his next class, but following the bell, a voice on the speaker said, "Aaron Anderson, report to the principal's office immediately." Aaron turned around and went to the office.

The waiting room was fairly small, with only one table for the receptionist and a few chairs. "Aaron Anderson," the receptionist said in a small voice. "They are ready."

"Thank you," answered Aaron. The receptionist did not reply. He went to the door with a window and a sign that read "Principal Cartwright." He went through the door. The room was fairly large, bigger than Aaron's bedroom, and had little furniture—just a desk and chair for the principal and two chairs next to one wall. Two men stood next to the room's only window. One of them, an average-height man in his thirties, had short red hair and round glasses, and he wore a blue Hawaiian shirt. The nametag on his shirt read "Principal Cartwright" in festive handwriting and had a drawing of a Hawaiian flower below his name. The other man had on a black suit, almost like a tuxedo. He was bald except for some gray hair on the sides of his head and he wore circular glasses. He also had a nametag, which read "Superintendent Gerald" in bold, formal type.

"Good afternoon, Aaron!" said Cartwright.

"Good, uh, afternoon," replied Aaron.

"The superintendent and I always like to welcome every student to our school," said Cartwright, in a joyful voice.

"Indeed," said Gerald, in a serious manner. "We want every student to feel...welcome at our school. Please, sit."

Aaron sat down in one of the chairs next to the wall. "Principal Cartwright, would you leave us for a moment?" asked Gerald.

"No problem!" said Cartwright. As he walked out the door, he waved to Aaron and said, "Good luck!"

Gerald then sat down behind Cartwright's desk. "So, Aaron," Gerald said in a slightly louder tone. "Tell me. How has your experience attending this school been so far?"

"Good," replied Aaron. "I have made a few friends, sort of, though they are all pretty strange, but I think we can all establish a relationship."

"Who are these students?" asked Gerald.

"Um, there's Jade and Ethan Winston, Crystal Hardy, and Josh Tempest. Josh looks like he's going to kill you, but he can be nice," replied Aaron.

Gerald stood up, walked toward the window, and stood there looking out at the cafeteria with his arms crossed behind him, then turned around. "It's good to see that you have been acquainted. I'm sure you will establish a much greater relationship with them over time," he said with a grim smirk. He went back to sit and said, "Well, Aaron, this was a nice talk. You may go back to class."

"Yeah, same here. I mean, like, nice talk, same here," mumbled Aaron. Something about Gerald was suspicious. He was secretive and grim all at once. Aaron walked out of the door in a fast motion. He wondered why the superintendent of all people wanted to have a "talk."

Chapter 4

The next day at school was even stranger. When he sat at lunch, nobody would talk to him. They all just sat there with their heads down on their crossed arms. Ethan was not there, but Josh was. "Crystal?" mumbled Aaron. "Jade? Josh?" Only Josh lifted his head. He didn't say a word and immediately put his head down again. Though Aaron warmed up a bit to Josh already, his stare still made him shudder. "Oh my God, what happened to you guys?" asked Aaron.

"Please go," whispered Josh.

"What...I...I thought we were friends," said Aaron.

"Go! Now!" yelled Josh.

Aaron stood there, confused. "I...I don't understand. You are the only thing close to friends I have. Everyone else is... just stupid jerks. You are the only friends I will ever make here."

Josh continued to bury his head into his crossed arms. Aaron just slowly walked away. The rest of the day none of them would speak a word, not even Ethan during English class to answer questions.

When the bell rang that ended the school day, Aaron walked over to Josh, who still looked down. "Josh, I know something's up. Please just tell me!" asked Aaron. Josh still looked down. Aaron never thought that he could build up the courage, yet he put his hands on Josh's shoulders. "Josh, please!" Aaron cried.

Josh lifted his head. "Meet me," Josh whispered. "Go to the park down the street from the school when the bell rings." Aaron took his hand off Josh's shoulders. Josh turned around and headed to his last class. Aaron was confused. *Park? Down the street? Today?*

After school, Aaron walked down the street. There were only a few houses, and most of them were run down. At the corner he saw a park with three swings, but only one was functional. The rest were just rusty chains hanging from a thin pipe. The only other feature was a slide, but it was severely cracked. This playground area was surrounded by sand, which was enclosed by a circle-shaped sidewalk. There was a rusted metal light post covered with graffiti, dents, and even a few bullet holes.

Aaron looked up and saw that the street light was not on. It was still daytime, of course, but Aaron noticed that the bulb was cracked. As he stared at it, a bluish-purple spark made the bulb light up. The spark dropped down the pole and onto the sidewalk. It crawled into a big drainage pipe next to the park. Aaron was so startled by the sight; he started to follow the spark. He approached the drainage pipe. It was almost like a tunnel. Aaron considered whether he was interested enough to follow the spark down to the sewer. His curious mind-set made him kneel down and start crawling before he could even answer his question.

The drain was actually fairly clean. Aaron expected bunches of moss and stains from water leaks, but it looked like someone had actually scrubbed the pipe. It was still dark, of course, and it took quite a lot of crawling to find the spark, but he caught up to it. He was face-to-face with the spark, which grew progressively brighter until Aaron had to look away. When he looked back, the spark was gone. Aaron crawled as fast as he could to catch it. In the dark, he didn't see the spot where the pipe suddenly bent down at a ninety-degree angle. His hands slipped on the edge, and he started free-falling straight down the pipe. He started to scream. Aaron noticed the spark falling right below him, and beyond the spark, he saw the lit bottom of the pipe. When he reached the bottom of the tunnel, the spark made a sharp ninety-degree turn. Aaron believed he was going to die, or at least break most of his bones when he hit the bottom. Instead of suffering a lethal hit on concrete, he landed on a pile of trash bags, which broke his fall.

Aaron was shaken, but he recovered quickly. The fall down the pipe seemed like a minute or two, but it was only a second. He looked around. He was in a room about the size of a large bedroom. He spotted a couch, a pool table, and a few doors. A figure stood up from behind the couch. It was Josh. He raised his hand to the height of his stomach. The spark was levitating on Josh's palm. He flicked his finger, and the spark disappeared.

CHAPTER 5

Josh put his hand in his pocket and pulled out a cigarette. He snapped his fingers near the butt, which made an electric spark that lit the cigarette. Aaron was amazed, but terrified. "Well…that's…shocking," Aaron mumbled. He gave a small chuckle.

Josh put the cigarette in his mouth and said, "Ha-ha," in a sarcastic tone. "You still can come up with a joke, even though you're terrified. Optimism is a good trait." Josh pulled the cigarette out of his mouth, tossed it on the ground, and stomped on it. "I would not normally waste something like that, but we have to get down to business," he said.

Josh knocked on the doors and said, "He's here. He's terrified. His legs are not broken." The door opened, and out came Crystal, Jade, and Ethan looking serious.

"I know this may be rushed, but we'll explain. It's time you knew," said Crystal.

"We don't even know if he's one of us yet," warned Ethan.

"Let's test that," said Josh. Josh tackled Aaron and started choking him. Aaron struggled for a few seconds, unable to push Josh off. Suddenly, Aaron felt a rush of adrenaline and shoved Josh away so hard that he crashed through the door. "Did that answer your question?" Josh said to Ethan as he brushed the wooden chips off of his lap.

Aaron looked at his hands. They were sealed in an orange, crystal-like armor. He stepped back and tripped. Without warning, an orange chain shot out from his hand and latched on to the pool table's leg. Aaron pulled on it gently and knocked the pool table over. Aaron looked down and saw that his whole body was covered in the same orange-crystal armor. Jade went to the other room and brought out a mirror for Aaron. He looked in it and saw an orange helmet. It had no eye holes or any holes to breathe through, yet he could breathe and see clearly. The helmet melted like hot metal into the rest of the armor and revealed Aaron's worried face. "Is this…a…wha…I'm…"

"You're an extraordinary," said Josh. "We all are." Josh looked to Ethan, then Jade, and then Crystal for some kind of agreement. They all nodded.

Josh stood still, like he was waiting for something to burst from his body. A black substance that shined like diamonds, but was like elastic, started to crawl up his legs and onto his shoulders. He skin started to turn to a dark shade of purple. His veins were light blue, and so was his hair. His pupils turned the same light-blue color and widened, filling both his eyes. The black substance formed a sleeveless armor from his feet up to his shoulders. He twitched his fingers and emitted a ball of lightning that floated on his palm. The ball of lightning shrunk and disappeared as if Josh had absorbed it into his hand. The black-crystal elastic armor started crawling off Josh's shoulders and melted into his feet with no trace. His pupils were black again, and they started to shrink, revealing his purple-brown irises. His hair slowly went back to the dark black it was before, and his skin turned to its normal pale shade. Aaron was dumbstruck. "How come…his skin… I did not—"

"We will explain after we show you. This is important for you to know," said Crystal.

She stood in the same stance Josh had before he changed. A clear elastic substance like Josh's emerged from Crystal's stomach, only hers was thicker, shinier, and pink, like her hair. It latched on to her skin, from her feet to the start of her hairline. Yet again, there were no eye holes in the helmet, but Aaron could tell Crystal could see just fine.

"Well, your name fits you!" Aaron shouted. Crystal stomped on the ground. Big, pink icicle-like spikes emerged from the ground and shielded her foot. On command, she turned back to normal.

Jade's armor was the same material, only blue, and it emerged from her hands. Her armor was sleeveless, like Josh's. It also wrapped around her mouth and nose like a bandit's bandana. Her pupils turned blue like her irises. In both hands were crescent-shaped blades that glowed. She showed them off by scraping the ground, leaving a massive crack in it. Then she too, turned back to normal, and Ethan stood up. *I wonder if they saved the best for last,* Aaron thought.

Ethan unfastened his prosthetic feet and jumped. His transformation was as quick as he jumped. The same substance, green this time, emerged from his shoulders and covered his whole body except his head. His already long nose stretched a bit and his green eyes became rabid. His neat blond hair became messy. From the ends of his legs emerged green, birdlike feet with three talons. The armor on his arms stretched out wide to form wings. He flapped his arms gently and started to hover. "I guess this is the reason I was born without feet," Ethan said with a chuckle. His voice was now hoarse like a screeching eagle instead of his regular gentle tone. He sat down and transformed back to normal. He fastened his prosthetic legs on again.

"How are we so different, but the same?" Aaron asked.

"Our bodies are made up of a shape-shifting material called extraordinarium," answered Ethan. "It makes up all of our bodies. No organs, no blood, just extraordinarium. It is passed down by rare genes, but the tricky part is that you are never half extraordinary or half human. You are either one or the other."

"Are there any more of us?" asked Aaron

"No," answered Ethan. "Our kind was considered a threat by the world, and all of them were exterminated by the EPA, except us."

"What about our powers? How come they are so different?" asked Aaron.

"Like I said, extraordinarium can change shape, color, and even the type of matter. Extraordinaries like us are born with special abilities granted by the extraordinarium in our bodies. When we transform, it emerges from us and changes us," said Ethan.

"You still didn't answer my question," said Aaron.

"Our different forms—you see the different shapes and armor we have? That's extraordinarium shape-shifting to match our ability. You are just one of the lucky few who can consciously control their shape-shifting powers. If you think it, you make it," answered Ethan.

"If I control it, then how did the freaking chain come out of my hand without me telling it to?" asked Aaron.

"There is only one other substance in extraordinarium—unlimited adrenaline. It boosts our strength, speed, and reflexes," said Ethan.

"I think I got it…but I'm still confused about our 'reputation' in the world."

"Cartwright will explain that," answered Ethan.

"Wait, the principal?" asked Aaron.

"The principal who is secretly on our side and training us," said Josh sarcastically, but with conviction. "OK, question time is over. Let's go," he said.

"How? I don't remember seeing a ladder on the way down," said Aaron.

"There are ridges and cracks in the concrete, but that's not how you're getting up," Josh said. He placed his hands on the sides of Aaron's head. Electricity pulsed through his arms and into Aaron's head. Aaron saw an electric portal beneath his feet that sucked him in.

Aaron closed his eyes. When he opened them, he expected to be falling in a dark ravine of despair, since it seemed like Josh hated him from the beginning, but he was back at the park. He looked around. There was no trace of anyone. Aaron went to sit down on the swing to collect his wits, but the swing broke. A mix of confusion and embarrassment crowded him as he sat there in the sand.

Chapter 6

Aaron kept his head down the next day at school to avoid suspicion. Aaron didn't know if someone was looking for him, but it was better to be safe than to be discovered.

The same receptionist was at the desk outside the principal's office. Without a word, she pointed at the door. Aaron opened it and found all his friends sitting on old wooden stools. Principal Cartwright was at his desk wearing his signature Hawaiian shirt. Josh was smoking, but Cartwright didn't seem to care.

"This is the second time I have seen you dressing casual," said Aaron.

"I'm not into the fancy stuff," Cartwright replied. He stood up and said, "Everyone, follow me to the truck."

They all went outside and walked into a massive navy-blue truck. There was only one seat in the very front, for the driver, who just focused on the road and ignored Aaron and the others. The rest of the truck was a large storage room with two long metal benches that could probably fit six people. Cartwright, Aaron, and the rest sat on the benches. Cartwright took out a tablet.

"Where are we going?" asked Aaron.

"I will answer that question when we get there," replied Cartwright. "But for now"—Cartwright closed his tablet—"I will explain a few things. This must be very strange for you, Aaron, but it is the truth. You are a part of an experiment designed to make the world see the true light of extraordinaries. Before I get into more detail, I think it is important that you know your friends' pasts." Cartwright stood up, leaning on the bench for balance since the car was moving. He pointed at Crystal. "Crystal was an orphan living on the streets. All she learned was to steal, beg, and hurt. When she discovered her powers, she started robbing and killing. She spent her life until now in a titanium cage guarded by the government."

He pointed to Ethan and Jade. "Brother and sister. They, too, were orphans, adopted by a troubled couple. Their foster mother reported them after Ethan clawed his abusive father's eyes out. Jade tried to stop her mother but ended up killing the woman with her uncontrollable blade. They were on the run for a year but were picked up by us."

Cartwright pointed to Josh. "You may be wondering why Josh can control electricity. We were stumped as well. It turns out that his extraordinarium has shape-shifted into a powerful variation which is very sensitive to energy and a stellar conductor. His unstoppable bursts of electricity killed his felon foster father."

Cartwright sat back down and looked at Aaron. "They are all the same. You are just lucky enough to not have a checkered past. Just trust us and we will help you." The tires screeched. "We're here." said Cartwright.

They all got out of the truck in front of a huge, cube-like building. It looked like an abandoned warehouse. It was placed in an ocean of dried grass. They walked toward the face of the building. Cartwright put his hand on a rusty metal wall and pushed. A bunch of dust fell when a door opened.

They went in, but the warehouse was engulfed in darkness. Cartwright banged on the wall and one by one, all the ceiling lights started to shine. The warehouse was empty. The light revealed a shining green button on the wall. *That must be the light switch*, Aaron thought.

Above was a lever, which Cartwright pulled. There was a lot of creaking. All of a sudden, objects started rising from the ground. There were metal poles stacked side to side, almost like a staircase. There were massive punching bags, each one about the height of a basketball star and the width of a sumo wrestler. On the ceiling, there were many turrets hanging from the ceiling painted red, but they had scratches that penetrated the coating. There were many other types of equipment, but Aaron had no idea what they could possibly be used for. "Consider this your elective that lasts the whole day," announced Cartwright.

He guided Ethan, Jade, Crystal, and Josh to various training stations. Cartwright led Aaron to the end of the building. Again he banged on the wall and a door opened. Inside was a small room with a metal chair in the middle.

"Sit down and close your eyes," demanded Cartwright. Aaron obeyed. "OK, now open." Aaron opened his eyes. Cartwright was pointing a pistol right between Aaron's eyes.

Chapter 7

Aaron felt a rush of adrenaline pulse through his body. He knocked the gun out of Cartwright's hand and almost punched him in the nose. He looked at his fist and saw his orange armor. He had once again unwillingly transformed into his extraordinary form.

"Excellent," said Cartwright. He put the gun down. "I just wanted to see your transformation. Now relax." Aaron obeyed and watched the crystal orange elastic crawl off his body toward his back. He again had his normal clothes on and no mask. Cartwright had a big smile, like he had just cured cancer. "Let's begin. Follow me to the next room."

They walked past Crystal, Ethan, Jade, and Josh, who were fighting a car-sized titanium mechanized scorpion dummy. Crystal head butted the dummy, which lost its balance.

Ethan was soaring through the large open warehouse. Some of the spikes on his wings darted from his arm and shot the dummy.

Josh had lightning spurting out from his hands, which made him fly. The middle of his chest blasted a powerful ball of electricity that froze the dummy for a while.

Jade climbed up the scorpion's body using her blue blades the way a mountain climber uses picks. She pushed one blade into the dummy's slit-like eyes and repeatedly stabbed its torso with her other blade. All five seemed a bit clumsy and unprofessional, but after a while, the dummy fell to the floor.

Aaron was amazed by his teammates' skills. Cartwright called Aaron's name. He was in the back of the building next to a door. Aaron had unknowingly frozen in place to watch the action. He ran back to Cartwright.

Cartwright put a firm grip on the doorknob and pulled. The door was locked. Cartwright's face brightened as if he had just gotten an idea. "Well, shape-shifter, first lesson. Make me a key."

Aaron focused on the doorknob's lock. *Think key, think key.* An orange key formed that levitated on the palm of his hand. Aaron flicked his finger and the key unlocked the door.

The room looked like an interrogation room. There was one rusted metal chair with handcuffs attached to the legs. Many different tools—buzz saws, knives, even a gun—hung on the wall. "Why do you have all these random rooms with chairs in them?" Aaron asked.

"Sit down," said Cartwright.

As soon as Aaron sat on the chair, the lights turned off and he felt the hand cuffs click. He was trapped. "Calm down," Aaron heard Cartwright's voice. Before Aaron had time to think, he heard the buzzing of a saw. He could see nothing in the darkness. He didn't know how close the saw was. He jerked back. He heard the buzzing scraping something. The lights turned on.

Cartwright was trying to saw a floating orange brick. The brick morphed into a baseball bat and struck Cartwright. He dropped the buzz saw and fell back against the wall. He sat up and clapped even though he had blood dripping from his nose. "Sorry. I saw danger, so I reacted," said Aaron. "May I undo the cuffs now?"

"Sure," answered Cartwright. Aaron formed a key and freed himself. "Break the wall," Cartwright said immediately after Aaron stood up.

"What?" Aaron asked, confused.

"Make something to break the wall. Then we will go on to our next lesson," answered Cartwright. Aaron tried to think of something creative, not just a wrecking ball, but maybe a giant sledgehammer!

He looked down and noticed something on his hand. His hand was gone! Instead, there was an orange spiked hammer that looked pretty sturdy. He examined it and punched the wall out of excitement. The entire wall in that room crumbled. There must have been at least four or five thick layers of concrete! Aaron's hand turned back to normal. He stepped outside. There were miles of dried grass.

"You can't get around much on your feet here, can you?" asked Cartwright. Aaron looked far into the horizon. He sensed that Cartwright had the same idea that he had himself. "Go on," he said. He sensed that Cartwright wanted him to travel, or drive a car, or even fly. He transformed into his extraordinary form.

Think travel; think fast; think, thought Aaron. He turned his neck. Cartwright was still there, sporting a friendly expression that almost shouted something like *I know you can do this!*

Aaron also sensed something protruding from his back. They were orange, but he couldn't tell what they were. "What's on my back?" asked Aaron. "Jet wings." Cartwright replied. Aaron smirked and then laughed. He pushed with his legs and, as his inner child hoped, he burst through the sky.

Aaron could almost feel the wind on the wings as if they were a part of him. He noticed a massive city next to the suburb where his house and school were. He waved his hand. *They can't see me, stupid,* he thought, but he was so overwhelmed with happiness that he didn't care. He did a 360 in the air before his wings folded back onto his suit. Aaron suddenly realized that his ultimate dream had become life threatening. He was now falling in the sky toward his suburb. All the buildings were growing as he fell.

Aaron pushed his hand out and formed a tower. He used orange claws to grip the tower, creating friction that slowed his fall. The rate of his fall decreased until his whole body hit the road.

He rested there for a minute until he gathered the strength to get up. Once he stood, he turned around and a car rammed him. To Aaron's surprise, he was still standing.

He stepped back. He got a closer look at the shiny blue vehicle. The airbag blew up and muffled the driver's scream. The front of the car had an imprint of Aaron's armor.

Sirens started to scream. Aaron looked up. There were two helicopters whirring around him with guns pointed at him. "This is the EPA!" the intercom on the helicopter yelled. "Move and we will fire!"

Chapter 8

Aaron raised his hands. "Remove your armor! Do not use your abilities to your advantage!" someone yelled over the intercom.

Aaron tried to command his suit to disappear, but it didn't work. He then realized that his powers wouldn't give in so easily. They were giving him a chance to run.

Aaron knew he couldn't fly because of the helicopters. Aaron thought as quickly as he could, and two large orange discs appeared on his ankles. Two words immediately came to his mind—*roller skates*! He turned and ran.

After a few steps, the discs started to spin, sending Aaron down the road at the speed of a race car.

Aaron heard the helicopters whir faster. He felt the energy of the bullets shooting right behind his feet.

Aaron looked at the road in front of him, which had started to fill with cars. He was heading toward the city. He saw the bridge up in front of him, followed by traffic. There was no way he could continue without attracting attention from civilians, so he tried something different.

Aaron formed drills on his hands and dove into the road. He mined through the asphalt until he got under the bridge. One of the helicopters hovered under the bridge and fired at Aaron. He formed a cannon on his hand and shot at the helicopter's blade. It swerved out of control while still firing, and accidentally shot the second helicopter, which also lost control. The helicopter under the bridge tried to fly up and over the bridge, but it hit the bridge with its blades and plunged into the water.

The other helicopter was now falling toward the gap in the bridge. Cars on the bridge backed up the other way as the helicopter plummeted into the water.

Fire was everywhere and the smell of oil filled the air. After a few seconds, the remains of the first helicopter exploded. Then the second exploded, causing a big wave of water to hit the bridge and completely destroy any remnants of it. Aaron quickly fled the scene before anyone could see him.

Chapter 9

"**What the hell** were you thinking?" demanded Josh. "We have nothing left. We have to live in a freaking hole in the sewer, and you try to take even more by exposing us?" Josh grabbed Aaron by his shirt and slammed him against the wall. Josh's eyes shined an angry-looking electric blue.

Aaron exclaimed, "Look, I'm sorry. It was self-defense! The sirens went off, and the helicopters—"

"This is my fault," said Cartwright. "I should have kept an eye on Aaron. I underestimated him. I had no idea he would fly so well and so far!" Cartwright chuckled.

Josh let go of Aaron. His eyes turned back to normal. "Why are you laughing? Are you some kind of sick f—"

"Oh, look at the time! I need to contact the driver," said Cartwright, tapping his watch. He pushed heavily on the door, and they all settled on the side of the warehouse until the truck came. "I didn't get the time to ask you this, but,"

Cartwright said, "what do you want to call this place?" The freshmen stood silent.

"Base," Jade said as she stood up. "We spend most of our time here. We are together. It's like a base." Everyone nodded.

"All right," said Cartwright. "From now on, each day we will go to Base."

They settled into separate conversations while waiting for the truck. Jade and Crystal started talking to each other. Cartwright was congratulating Ethan on his training, but Ethan countered by saying he was still a bit rusty. Aaron tried to absorb it all carefully. He still couldn't quite believe his situation. Josh stood by himself. He plucked a dried piece of grass, which he electrocuted. A flame burst from the tip. He took out a cigar and lit it with the piece of grass. He pinched the piece of grass with his fingers and dropped it. A little electric spark crawled quickly from his skin back into his fingertips. Aaron felt a bit sorry for these guys; they were all orphans living in the sewer.

Aaron slowly tiptoed over to Josh. He tried to think of something that would break the ice. Instead, his mouth spilled out, "Why do you smoke so much?" Aaron curled his lips inward before he could say anything else.

Josh looked at him awkwardly and then sighed. "My dad did it a lot. He always had empty smoke boxes lying around the house. One day, his drunken ass left a fresh one open on the ground; I got into them, and I haven't stopped since."

"That stuff hurts you—or do you not care?" asked Aaron.

"They don't hurt me. One thing my powers do is heal me from injuries faster than normal. My lungs, or whatever extaordinaries use for lungs, just repair themselves. The addiction is real, though."

Aaron jumped up when he heard the honking of a truck; he was desperate to get home. Aaron finally thought of something smart to say to Josh. "Just remember that even if it doesn't affect you, it will still affect the way people think of you."

Josh seemed to get the message. Josh pinched the end of the cigarette and stomped on it before he got into the truck.

"How did they do this time?"

"Better than expected. They all have their own talents."

"What about *interference*?

Heard he made quite a ruckus out there."

"Indeed. Are you sure we can't notify the staff about everything that is going on with this mission? They will be better suited to prevent events like this again."

"Never. That puts us at too great a risk.

This was meant to be a secret project, and it will stay that way until the kids are ready.

We will prevent the agents and the kids from asking any questions about their own missions. Just keep me updated."

"I have discovered that they are all emotionally weak.

They trust us. We can use that to our advantage."

"They are all weak. We need to fix that. They must become strong, yet still oblivious.

Our only risk right now is *interference*."

"Due to his past being not as bad as the others',

He doesn't seem as breakable as them."

"Exactly. We need to eliminate someone close to him.

We need to make him feel weak and alone. That is sure to break him."

"Sir, do you have faith that this will work?"

"Yes. It has to. Even if it fails and they turn on us, we will control them with fear.

Project G is sure to work. Do not contact me until everything is finished."

ᛪ

OK, so what will I say? Aaron thought. *Oh, hi, Mom! I was just at school learning stuff and things…*No. *Hey, Mom! You're the best! I'd rather not talk about my day, so good night!* But he had gone to bed early on the first night of school. Doing so twice would be too suspicious.

Before Aaron had any extra thinking time, he arrived at his front door. Aaron heard Snowball bark and scratch on the door.

Aaron's mom opened the door. "Hey, honey! I didn't expect you to arrive so late! How much are they torturing you there?" she said with a chuckle.

"Everything's good," Aaron replied. He stepped inside. "I am going to watch some TV before I get to my homewo—" Aaron had no homework. He didn't know he had any, at least. "Uh, actually the teachers gave us a break; I don't have any homework," Aaron said.

"You are very lucky!" said Deborah. "Most kids in high school have lots and lots of homework by now, and they don't get a break."

"Yeah, Mom, I think those loads of homework will come tomorrow," replied Aaron.

Aaron jumped into his favorite chair, which he had claimed since his dad supposedly disappeared. The brown leather on the recliner was ripped and had cotton sticking out, which Snowball liked to grab and make a nest with. The dog sat on Aaron's lap, chewing a piece of cotton from the chair.

Aaron turned on the TV and zoomed through the channels until big red letters that spelled EPA popped up on the screen. "An extraordinary being has been spotted!" the TV repeated. The camera zoomed in on an anchorman who looked like the baby of skinny and ugly. "Breaking news!" he shouted. The TV showed footage of Aaron from earlier that day. "A major terrorist attack by an extraordinary occurred this afternoon on the main bridge to Star City from the Gilbert district. He took down two helicopters and killed several—"

Wait. Did I kill anyone? The TV showed bloody bodies on either side of him. Aaron was sure those were not there. There were even added explosions in the background when the helicopters crashed.

Aaron threw the remote on to the couch beside him. He was frustrated even more about the world's hatred for extraordinaries since now knew he was one of them. Snowball jumped off Aaron's lap, and Aaron stood up. He went to his room and tried to calm his anger.

Chapter 10

Aaron had a long, sleepless night. He lay in his bed wondering about the next day. *Am I going to disappoint at training tomorrow?* He thought. *Will I train more, or will I wake up and go back to my ordinary life?* Aaron shuddered at the thought. If this was real, there would be no such thing as an ordinary life anymore.

He stared at the ceiling, wishing it would stare back, noticing his problems. Snowball was sleeping next to Aaron's legs, but the dog suddenly jumped off the bed and started scratching the door. Aaron sprang from his bed and headed for the door.

As soon as he touched the door knob, he heard a series of bangs, screams, and then silence. Aaron jerked the door open and ran through the house. Snowball ran at Aaron's pace. Aaron opened the door to his mom's room.

Everything was still. The window was broken, and glass was scattered all over the floor. In the elegant queen-sized bed lay the bloody body of Deborah Anderson. Aaron ran to the side of the bed, staring at his mother's corpse. Snowball curled up into a ball on the polished wooden floor, which was now stained with blood. "Mom!" He whispered quietly.

Aaron felt a tear rolling down his cheek. He hugged his dead mother, hoping that some miracle would resurrect her, but the miracle was absent. He sat by his mother's side for a good five minutes until he stood and looked out the broken glass window.

He saw a light glimmering in the night sky. It looked like a shooting star moving fairly quickly. It looked as if it was…a helicopter. Aaron thought maybe the helicopter had something to do with his mother's death. *Fat chance,* Aaron thought, but it was worth checking it out.

Aaron jumped out the broken window and slid on the shingles on the roof, cracking a few. When he reached the edge of the roof, he jumped with all his might. He changed to his extraordinary form and flew into the night sky. His armor protected his face from the piercing wind. Aaron caught up to the helicopter and froze; it was only a news copter. The front lights that were directed toward the city below now turned on Aaron. Aaron tried flying away as fast as he could, but the lights followed him. "The helicopter maniac! Fly away!" a man in the helicopter shouted. Aaron created an arrowhead and threw it toward the light like a playing card. Several sparks flew from the bulb and the light went out. The helicopter virtually disappeared.

Aaron looked back at his house and darted toward it. He was filled with sadness and rage—he wanted to find his mom's killer! For now, Aaron could do nothing but scream in emotional agony. Aaron reached his house and slowly floated through the same window he fled from.

He changed back to his normal form and once again mourned over his mom's corpse. He reached his hand toward his mother's and grasped it. He wanted to cry again, but it was as if he used up all his tears.

"Sir, I know you don't want to hear from me until Project E is finished, but I just want to notify you that Subproject Elimination was a success.

"Good. Make sure the police do not follow the case.

When morning comes, house *interference* with the others.

We can't have him living in the house.

That would mean too much interaction with the outside world."

"Yes, sir. Also, there may be a delay in Project G due to *interference* being discovered. The media's calling him the

Helicopter Maniac. If we don't stop it, we'll be exposed."

"We shouldn't worry about it. Not even the media will find out about us.

Just keep on working with the subjects. If there are any problems, I'll handle them."

Chapter 11

Aaron ran to pick up the phone. He dialed 911 and awaited an answer. A middle-aged-sounding lady answered, "This is Star City Police Department, what is your emergency?"

"Yes, I was—" Aaron had to catch his breath. "I was in bed and I—" He had to take another deep breath. "I heard a scream, so I ran to my mom's room, and I—" The crackling sound of interference filled the phone, loud enough for Aaron to pull the phone away. The interference had stopped. Aaron tried pushing the buttons again, but it was if the phone had lost power.

Snowball was barking loudly. Aaron shushed him and then thought of the impossible—he needed to go to the police headquarters in the city. Aaron rushed out the door and thought about flying there. *I have already caused enough trouble in the skies. I will be spotted. Plus, I destroyed the only bridge. How am I going to explain?* He thought. *I have no choice.*

He went into his extraordinary form and burst into the air, soaring with his wings. He could feel the cold night air, despite being covered in armor. It soothed him a bit and took his mind off his mom's death. He flew over the large gap between the Gilbert suburb and the city.

He flew into a dark alley and turned back into his normal form. He crashed into a garbage can. He got up and reached the police station and burst through the door.

His ruckus made the receptionist stare at him. "May I help you, young man?" she asked with a bitter tone.

"My phone stopped working; I couldn't call the police department, so I ran here," Aaron replied.

"In which suburb district do you live?" she asked.

"Uh, Gilbert," Aaron replied.

The receptionist leaned on her desk and stared at Aaron. "So you are saying you came here from the only bridge that connects our City to Gilbert, which has been completely destroyed?"

Aaron knew this one was coming. "I'm an excellent swimmer; please don't ask any more questions. So can you help me?"

The entrance opened once again. Cartwright stood hunched and out of breath. Aaron's eyes widened. He did not expect Cartwright of all people to be here. "I am sorry, ma'am," he panted. "My son here is taking hallucinatory medicine and has a crazy imagination on top of it all." Aaron could see that Cartwright did not expect the receptionist to believe his story. "We live in an apartment down the road."

The receptionist stared at Cartwright for a second and then said, "Whatever. Just look after your son." Aaron heard her murmur something like, "I'm too old for this job."

Cartwright and Aaron left the station. Cartwright put his hands on Aaron's shoulders. "I know what happened," said Cartwright. "The police cannot help with this. My team and I are working on finding the killer. What I want you to do is get some rest. I will give you a ride home using the ferry, and I will pick you up tomorrow from the school at seven." Cartwright patted Aaron.

"What about my mom's body? You can't just leave it there." said Aaron.

"We already disposed of the body. Don't tell anyone except your teammates," replied Cartwright.

Aaron walked with Cartwright to the docking station at the boardwalk. Aaron wondered why he didn't just create a boat. That would have been more believable than what he had already said. Aaron got home and looked at his mom's bed. No body lay there. He once again cried next to the bedside, wishing she was still there.

Chapter 12

The next day, Aaron got to school as usual. When he arrived, he saw Cartwright, Ethan, Josh, Crystal, and Jade at the entrance. They started loading into the same truck as yesterday. Aaron got into the truck and sat next to Ethan and Cartwright. His friends gave him sympathetic looks signifying that they knew about Aaron's loss.

Aaron could feel the immense speed of the truck and the sharp turns throughout the journey. Within a matter of seconds, they unloaded from the truck.

When he got out, Aaron saw the same park he had walked to the other day. They headed straight for the drain pipe and down into the hideout.

Nothing had changed from before except the three doors now had labels on them and the door and pool table were now fixed. The first door read "Bathroom," the second read "Jade and Crystal," and the third read "Ethan, Josh, and Aaron."

Cartwright opened the third door. There were three beds, one for each boy, and a small closet with clothes for each of them. There also contained a dresser for each boy.

"Well, I have a lot of work to do, so you can just take this day to relax. Oh, and one more thing," added Cartwright. He went into the boys' closet and pulled out a small kennel with Snowball inside.

"Snowball!" yelled Aaron. Aaron accidently tore the door apart to the kennel with a big orange claw. Cartwright took a step back in shock. Aaron hugged Snowball and set him down.

Josh walked up to Cartwright and put hands on Cartwright's head. "I hope I survive this," said Cartwright. Blue electrical pulses went through Josh's arms and into Cartwright's head. Cartwright made a small cry of pain before he disappeared into a vortex of electricity that quickly closed.

"Why don't you just teleport us everywhere all the time?" asked Aaron.

Josh sank down into the couch. "It takes a lot of energy," answered Josh. He immediately took a cigarette out and started to smoke it. They all sat down and talked for a while.

"Now that we are all settled, I think we need to come up with code names based on our abilities," suggested Jade.

"Crystal already fits for me," Crystal answered with a small chuckle.

"How about Vulture for me?" asked Ethan.

"Nah. It makes you sound like a villain," said Aaron.

Ethan scratched his head. "How about Eagle?"

"It fits," answered Crystal.

"Eagle it is," confirmed Ethan.

Jade summoned her blue moon blade. She inspected it. "Hey! How about Moon Blade?"

"Very clever," Crystal sarcastically remarked. "How about you, Aaron?"

Aaron felt a bit awkward. He never would have imagined sitting in a group of fellow freaks trying to come up with nicknames.

"I can shape-shift," said Aaron. "Shape-Shifter would be weird, though."

"How about Shift for short?" suggested Ethan.

Aaron nodded rapidly. "I'll take it. What about you, Josh?"

Josh took out his cigarette and just silently stared at Aaron.

"I have an idea," said Aaron. "I remembered you giving me a shock when I first saw you." Aaron lightly chuckled.

Josh stood up and gave Aaron a death stare; then he grinned and said, "Accepted." He sat down and resumed his smoking.

"All right, Shock," said Aaron. "The last extraordinaries on Earth are officially united."

A silence followed, and then Ethan sighed. "What's the matter?" asked Aaron.

Ethan had trouble finding the words. "I...I don't want to live in this hole forever, hiding from the real world. I hate being hated, you know?" Aaron felt compassion for him, but then he had an idea.

"Guys?" asked Aaron. All attention was on Aaron. "Do you want to help me find my mom's killer?"

Chapter 13

"**W**hat?" said Crystal sternly.

"Cartwright gave us specific directions—do *not*, under any circumstances, get involved," said Ethan.

"Hear me out, guys!" cried Aaron. "We are *not* going to kill him."

"Or her," added Jade.

"Thank you, Jade, for hearing me out," said Aaron. "I just want to see him, or her, brought to justice."

"What's in it for us?" asked Josh.

"I was getting to that," answered Aaron. "All of you say that you don't want to live in this hole. You want to be free. You want to explore the world. You don't want to hide. Well, number one, if we bring a criminal to justice, it will make us look good, and we may not have to hide. Number two, you don't want to live in the sewer, and I don't either, so all of us will live in my house."

"Exactly how will that work out? We can't just claim it as our own. I doubt Cartwright will allow us," said Crystal.

"When he sees our success, we can convince him," answered Aaron. "So will you guys join me?"

They all just stared at Aaron for a bit; then the person Aaron thought was the least likely to join spoke up.

"I'm in," announced Josh.

"What?" said Ethan.

"You can't be serious!" said Jade.

"There is no way we will pull this off!" said Crystal.

"We will if we work as a team," said Aaron.

"If that means that I get to walk on the streets without feeling hatred around me, I'm in," declared Josh.

Ethan stood up. "Josh is right. We can't stay here forever," he said.

Jade stood with her brother. "Are you coming, Crystal?" she asked.

Crystal just stared with doubt. "We are going to get ourselves killed, but I guess it may be worth a shot," Crystal replied.

"We start planning today and every day after training," said Aaron. "We need to find out who was on the other end of my phone besides the operator and me." They all sat and started to discuss plans.

ᕯ

"Sir, this is urgent."

"What now?"

"*Interference* is teaming with the others to find the killer.

They are making plans to infiltrate the police headquarters to trace the blocked call."

"Let them."

"Sir, you can't be serious."

"This will speed up the time it takes to reach
their potential."

"What happens when they discover that
we did it?"

"By then, Project G will be done. We can use
it to control them through fear.

Keep me updated."

Chapter 14

Aaron and his friends had discussed the plan for days, and now, Aaron had a plan. "In the police HQ, there's a room where they have records of every call. A computer there will have the information we need to find out who blocked my call," explained Aaron. "I'll get to the computer, but I need your help. Crystal, there's security at the entrance. You need to distract the guards and the receptionist while I find the room. Ethan will remain outside to listen for my signal. Ethan will tell Josh, who will be next to the building's power box, to absorb all the electricity to cause a blackout in the building. All the cameras will be down and all the doors will be unlocked. I'll find the room and signal Ethan to tell Josh to turn the power back on. With the computer on, I'll be able to find my call. Once I'm finished, Josh will turn the power back on while we retreat to the sewer."

"And what am I going to be doing?" asked Jade.

"I was getting to that," replied Aaron. "You will make sure the people outside are not suspicious. If anyone tries to get in the building, you need to distract them. Are we clear?" They all nodded. "Then let's get to work."

All five used the new public ferry since the bridge was gone. They went to the police headquarters and scouted. Aaron reviewed the plan, and then they all got into their positions.

Josh stood behind the HQ building, smoking a cigarette. Ethan used a ladder Aaron created to climb to the roof of a two-story building next to the headquarters, watching Josh and awaiting Aaron's signal. Jade stood on the sidewalk, scouting for any trespassers. Crystal and Aaron walked inside the building.

Crystal started speaking with the receptionist, making up a story as she talked. "So I was at my house, which is in the Palm district, and this guy just decided to pick the lock to my door." She continued to ramble like that at the slowest pace possible.

Aaron, who was standing near the front door, tapped the glass twice and knocked it with his fist. As soon as Ethan heard this using his heightened hearing, he picked up a pile of sand from the messy roof and threw it at the back wall of the police station. Josh saw the dust fly everywhere and he swung an electric pulsing fist at the power box. He then began absorbing the electricity from it, which made the police building and a few buildings near it lose power. Darkness flooded police HQ.

The dim light of dusk almost revealed Aaron sneaking behind the front desk into the main building and up the stairs. As he crawled, he could barely see the people shuffling downstairs. He managed to find a building directory, which led him to a room with an aluminum sign that said "Call Collective." He opened the door and found another old, rusty vault door with five locks still intact. Aaron used his training and formed five keys that unlocked the five locks. He opened the door, whose hinges screeched and squeaked.

The room had several computers and a single monitor next to a small window. This time, Aaron knocked the window only once.

In response Ethan threw another pile of sand to signal Josh, only this time he accidentally hit Josh. Josh coughed and looked back at Ethan with his signature death stare. An electric bolt zapped through Josh's arm and into the power box, which lit the building again. Josh aimed his other finger and sent a small spark toward Ethan's leg. Ethan grasped his leg in pain, hopping on his one prosthetic foot. He lost his balance and fell back onto the roof.

A man spied this exchange from the street with his face filled with confusion. Jade spotted this and quickly grabbed a littered napkin and a pen she had in her pocket and approached the man. "Hello sir, would you like to take a quick surv—"

"No, no, please don't," he replied, lightly jogging down the road. Jade's trick had erased the man's suspicion.

Meanwhile, Aaron got on the computer, but was blocked by a password. He started to search for any clues for the password. He found a convenient sticky note behind the monitor that read "122200137." He typed it in, and it worked. The screen showed several titles with the date of the call and the location. He found his file and clicked it. It showed a map of the whole city. Three pinpoints appeared on the screen. One was on Aaron's house, one was on the police station, and the third was in the ocean far from the city. He zoomed in and only saw the ocean's blue water. Aaron was confused; he expected to find the place where his call was blocked. Or maybe that was where the call was blocked. He wrote down the ocean coordinates on the sticky note that had the password. He put it in his pocket and was about to knock on the window when a guard opened the door.

Before the guard saw him, Aaron jumped up, turned his hands and feet into magnetic claws, and grabbed onto the ceiling. He hung there watching the guard type on the keyboard. The password screen showed up and the guard looked behind the monitor. He scratched his head and went back through the door. Aaron took one of his hands and made a cannon that shot an orange pellet the size of a gumball at the window. Once again, Ethan signaled Josh to absorb the power. Once it was dark again, Aaron crawled on the ceiling and then zoomed back downstairs.

He heard Crystal still talking. "See? He is probably the one that took the lights out!" she said.

"The act is over. We need to go!" Aaron whispered. The receptionist was glad that Crystal had finally stopped talking

By this time it was night so Aaron and Crystal could easily sneak out the front door. Everyone was gathered at the back of the building. They formed a circle and held hands. Electricity surged through Josh and the others until an electrical vortex sucked them all back into their living quarters underground.

CHAPTER 15

Everyone was amazed; the five never thought they would be successful. While the others cheered, Aaron sat quietly, overwhelmed with surprise and relief. After a few minutes, Ethan, Crystal, and Jade started playing a game of pool while Josh sat on the couch smoking a cigarette. Aaron opened the door to his room. There was a package on his bed along with a note. The note read:

AARON,
MY TEAM SEARCHED THE HOUSE AND FOUND A FEW THINGS YOU MIGHT WANT AS KEEPSAKES. IF YOU GET HUNGRY, THE OTHERS WILL SHOW YOU WHERE THE FRIDGE IS. BE READY TOMORROW BY EIGHT O'CLOCK. KEEP WORKING HARD!
—CARTWRIGHT

Aaron looked through the doorway and saw Ethan knock on the wall. A secret door slid into the wall and revealed a fridge. Inside the fridge were several trays of food. He laid the note on his dresser and opened the package. Inside were simple personal items—a toothbrush, toothpaste, and a comb.

There was also a picture of his mother and him. He almost shed a tear at the sudden sight of his mother. The last object was his mother's laptop. All of a sudden, he realized he could use the laptop to search the coordinates for more information.

He turned on the computer and punched in the coordinates. What popped up was a link to a website about the Star City Underwater Canyon. It said that it was supposedly haunted and that all who have swum in the canyon have never returned. Aaron knew there was something there that he needed to explore. He picked up the laptop and gathered his friends. "Guys, I've discovered something critical to our mission."

A

"They call it the Star City Underwater Canyon. Somehow, somewhere in that canyon, someone has built some sort of base," explained Aaron. "We need to figure out what is down there and how to get there."

"You can't expect that after only about a week of training we are ready to go down in a freaking haunted canyon and fight who knows what!" argued Josh.

Aaron replied, "I understand we still got some shaping up to do, and I still have to research this thing, but in the meantime, I think it would be smart to begin developing a plan!"

Aaron saw the stunned faces of his friends and realized he had been shouting. In a quieter tone, he said, "Guys, knowing that my mom died unavenged would be the worst thing in the world to me. I say we just train as hard as we can, learn what we need, and prepare ourselves for the worst."

After Aaron gave his speech, he left the room and plopped himself onto his bed. It didn't feel like the bed at home. This bed felt springy and dull. The thought of his house above the sewer made him shed a tear. Sadness morphed into anger. As his eyes flowed with tears, he stood up and jerked his hand to the side. He noticed that he had unconsciously slashed the wooden legs of his dresser with a long orange blade that replaced his hand. That is when he finally realized what he really wanted—he wanted to kill the man that killed his mother. That made him just as evil as his mother's killer. He was a monster. That is why society exiled him from having a normal life. He laid on his bed, covering his face the best he could. He eventually cried himself to sleep.

Chapter 16

Aaron woke long before the others. His back felt slightly strained from the uncomfortable bed. To distract himself, he ate a tray of bread slices and sausages for breakfast while he played a game of pool by himself.

As Aaron aimed the pool stick, he caught a glimpse of Josh on the couch and jumped; he missed the cue ball entirely. On Aaron's shoulders were orange missiles ready to shoot.

"Calm down, you pansy," said Josh. Aaron turned back to normal and approached Josh. "I just had some trouble sleeping. Don't worry, though. I don't get that much sleep anyway," said Josh. Aaron kept silent until the others started to wake up.

At eight, they climbed up the sewer pipe. The truck was awaiting them. They got to base and started their training. They mostly trained as a team battling the same scorpion dummy as before. This was Aaron's first time facing the beast. Instead of having several eyes like a normal scorpion, this one had three blue vertical slits on its face.

He heard the cries of his name, and he realized that he got distracted looking at the scorpion and was about to get stung by the scorpion's double-bladed stinger. He thrust the stinger away with his armored forearm. The scorpion grabbed Aaron with its claw and wouldn't let go. Ethan, or Eagle, swooped in and grabbed Aaron by the shoulders with his green talons, like a real eagle snatching its prey.

"Drop me. I have an idea," said Aaron, or Shift.

"All right," said Eagle in his raspy voice but still as polite as his other identity, Ethan. Eagle dropped Shift. Shift transformed himself into a spiked torpedo about to drop on the scorpion's torso.

Meanwhile, Crystal pulled on the scorpion's leg. Her pink diamond armor grew harder and shinier as she tested her might.

Jade, or Moon Blade, decided to climb up the scorpion's stinger and try to cut it off. The scorpion swung its tail, leaving Moon Blade hanging by one blade that was pierced into its tail.

Josh in his purple-skinned Shock form was electrifying the scorpion's claw, trying to heat it up enough to break it off. The metal started to glow orange, like Shift's armor, so Shock pulled and broke the scorpion's claw, exposing a few wires. He grabbed them and started to absorb the electricity.

Shift jumped into the air and then dived in like a bomb. His torpedo body hit the scorpion, impaled it, and caused it to break in two. The battle was done. Cartwright was filled with awe. He started clapping. "Excellent! You have all done perfectly!" He pulled a remote from his pocket. "Let's kick it up a notch, shall we?"

Cartwright pushed a button on his remote. The broken wires of the scorpion started to reconnect one by one, mending themselves. The outer shell began to attract other metal objects in the room like a magnet. The scorpion was somehow repairing itself. When it was finished, it stood up, much bigger than before. Its new forked stinger spun around and aimed at Eagle.

The scorpion shot the whirring stinger at him. Shift flew in front of Eagle with his orange jet wings and formed a shield around the two of them. The stinger bounced off the shield. Shift grabbed the stinger, created a giant orange slingshot, and shot at the scorpion. The stinger only made a scratch on its titanium armor. The scorpion retrieved its stinger by magnetizing it back to its tail.

"You have a plan, right Aaron?" asked Eagle.

"I think so," answered Shift. "We each need to attack a specific part. I will get the stinger. You get its body. Tell Josh and Crystal to get its legs, and tell Jade to get its claws."

Eagle flew around and spread the word. Shock electrified the scorpion's left legs, making them magnetize each other. The scorpion tried to escape by lifting its legs.

Crystal pounded on the ground near the scorpion's right legs. The ground toppled, causing the scorpion's side to fall into the crevice.

Eagle tried to pierce the scorpion's armor by throwing green shard-like projectiles from his wings.

Moon Blade gripped one of the scorpion's claws with both her blades. She pushed on the socket of the claw with her feet, attempting to rip it off.

Shift lassoed the stinger with an orange chain. He pulled the chain and flew over and under the scorpion's body and legs.

"Aaron! I broke its armor!" cried Eagle.

Shift pulled on the chain, tripping the scorpion. He formed a spinning drill on his hand and flew into the cavity in the scorpion's armor. Shift pierced the body, which made wires shoot sparks all over him. Despite being protected by his armor, he felt some stings from the electricity. He flew faster and returned from inside the scorpion.

The scorpion took a few wobbly steps before it fell. All five turned back to normal. They violently huffed, trying to regain their breath.

"You are all doing excellently," said Cartwright in a calmer tone than expected, but with a big smile on his face, like he was trying to conceal his excitement. Everyone gathered together. They shook hands and cheered.

Aaron approached the lifeless dummy. He approached the face. Its eyes no longer shined. Aaron cautiously knocked on the titanium shell. The scorpion's eyes started to glow red. It grabbed Aaron with its claw. Aaron transformed. He tried to squirm, but he could not free himself. The scorpion pointed its stinger right between Aaron's eyes. Aaron formed a shield around his head just before the scorpion pulled its stinger back and struck. The spinning stinger broke through

Aaron's shield. Aaron physically felt pain after this, like he had broken a bone. He scrunched his eyes, waiting for the next blow, but instead of being killed, he felt scorpion's armor starting to shed.

The dummy had unwittingly stabbed itself with its stinger and claws. Wires and chunks of metal flew around until the scorpion's eyes dimmed and went black. It fell to the floor, which caused the rest of its body to shatter. Aaron's friends had rushed toward him, asking if he was OK.

"I'm fine. It just scared me a bit," replied Aaron.

"I am so sorry," said Cartwright. "There must have been some bug in its processor. Until it's fixed, I will train all of you individually." They were unable to continue training, for the school day had ended.

They all loaded up in the truck. Cartwright was still apologizing to Aaron about the malfunction. "I am truly sorry and I hope you can accept my apology," pleaded Cartwright.

"Yes. OK. Affirmative! For the last time, you are forgiven!" Aaron begged. A moment of silence fell between them.

"Aaron?" asked Cartwright.

"Yes?" replied Aaron.

"Thank you. Thank you for cooperating and keeping your spirits high despite all you have been through," said Cartwright.

"And…thank you for being my mentor," said Aaron. Aaron had felt that he established a strong connection with Cartwright, like the father figure he never had.

"Sir, I have exciting news."

"Continue."

"All subjects are showing very impressive results.

They will be ready in a week at the most and so will Project G."

"Excellent. How is *interference* doing?"

"Each day he improves, but I fear he may be capable of destroying Project G."

"Yet he was helpless when he was trapped in its claws?"

"True, sir. What I meant to say is that he could greatly improve within a few days.

His strength might match the technology of Project G."

"Do not be absurd! This will work! If it doesn't, we can rip the damn stuff from their bodies.

Either way, we will be invincible."

Chapter 17

For the next several days, Aaron and his team trained and grew stronger each day. When the end of the week drew near, Aaron gathered everyone to explain his finished plan.

"Guys, I'm aware we don't know what is in that canyon, but we have taken the time to prepare, and I think we're ready," said Aaron. "We're going to walk to the lake at night like our normal selves. When no one is looking, I will create some sort of marine vehicle and transport us to the bottom of the canyon. There has to be something down there. Once we get inside, if you get attacked, attack them—or it, but our main objective is to find where they blocked my call. We have to find someplace where there would be technology advanced enough to block the call. The answers will be there." Aaron finished and told his team to prepare for the journey.

While he waited for his friends to get ready, Aaron rested in his room. He picked up the picture of his mom and him. He hugged it. "Your death will not be forgotten."

Jade knocked and entered the room. "Hey. How are you?" she asked.

"I am excited, but nervous," Aaron replied.

"We all are." She gave Aaron a kiss on the cheek. "When this mission is over, we will be able to do anything we have ever wanted," she said. Aaron did not expect such a thing coming from Jade. He thanked Jade for her comfort; then they went out to meet the others, and the five of them climbed up the sewer pipe.

⋏

When they reached the docks, there were several people around. Ethan pointed out a dark spot at the entrance of a cave. They walked over and then transformed.

"This is super ridiculous, but it should work," said Shift. He created a large orange bubble around his teammates and himself. It lifted off the ground, hovered in the air, and then slowly plunged into the water.

The bubble moved with the current in the direction of the canyon. "You were right," said Shock. "This is way too damn ridiculous to be working."

"A lot of efficient inventions are ridiculous," replied Shift. The orange bubble was fairly transparent. Everyone could see the dark light of the moon shining on the sand below until they got lower. Mostly bottom feeders roamed the tan surface.

After a half hour, they reached the dark ravine. The bubble dived in. "Light, please," said Shift. Josh electrified his hand, which made it glow blue and provided enough light to see the outside. Strange sea creatures crawled away and hid in crevices. Shift dived deeper. Shock, Moon Blade, Crystal, and Eagle covered their ears when the pressure heightened. Shift's shape-shifting ability allowed him to adapt to the pressure more easily.

They finally reached the bottom of the canyon, but they found only sand and seashells. "No," Shift whispered. "There has to be something here! I know it!"

"It's OK. We'll search elsewhere," suggested Moon Blade.

"No! It was here!" Shift exclaimed. His mask melted, revealing Aaron's worried face.

"Wait!" said Eagle. "There is something beneath the sand!"

The mild current made sand drift off, revealing a piece of metal. Shift stood up and formed an orange hand that grew from the surface of the bubble. The hand dusted off more sand. A much larger piece of metal revealed bold letters that read "EPA." Shift banged on the metal with the orange hand. The sand started to sink inward. A large black sinkhole formed in the sandy surface, and the bubble was sucked in. Shift tried to steer upward, but the suction was too strong. They were engulfed in a dark space. There was no noise until Shift heard water starting to drain. Josh glowed his hand again. The water was draining as Shift suspected. When all the water was gone, bright lights shone from above. Sirens started to shriek.

They heard a voice on the intercom repeatedly say, "Unidentified ship has docked!" Three guards came rushing in carrying sleek black rifles and other firearms. They examined the entire room, but found nothing.

"You two tell everyone that it was a false alarm. I will stay here just in case," said the lead guard. The two guards left the room and the lead guard protected the entrance.

Shift had attached himself to the ceiling with hooked hands. He had also made chains that hung from the ceiling and held his teammates by their waists. Shock made a gesture to Shift—a request to take out the guard. Shift nodded. Shift made the chain melt off Shock's waist. Shock fell onto the guard's shoulders and placed his electrified hand on the side of the guard's face. The guard squirmed and yelped and then fell to the floor.

Shift melted all the chains, and the five fell to the ground. As they looked around, they realized how big the docking station actually was. A truck could easily fit inside. They opened the door and found an even bigger hangar.

They were on the third floor in a large open area, looking over a gate, when they saw a massive metal scorpion the bottom floor. It was identical to the one they had trained with, but this one would probably be too big to be the scorpion they trained with. There were people walking all over, typing on computers, carrying folders, and some who were on the bottom floor working on the scorpion. On the bottom floor, a large door opened on the side of the room. It contained a massive submarine, even bigger than the scorpion. Workers loaded the scorpion into the submarine. The door closed and the submarine left. Everyone started to cheer. Workers in lab coats shook hands and guards threw their hats in the air. They celebrated until one guard pointed out Eagle's shadow.

"What is tha...wait. Those things are here!" the guard yelled. "Those rotten things are here!" The scientists fled to their labs and the guards hoisted their firearms. "You guys can take them on! I will find the communications room," said Shift.

Eagle flew around, grabbing guards from everywhere with his talons and releasing them from high up.

Crystal dashed toward a guard and pounded him into the ground, which made the ground break, sending them falling to the second floor.

Moon Blade jumped and fell on a guard. She whacked his head with a double fist that knocked the guard out.

Shock was shooting at guards with electric sparks spurting out of his fingers.

Shift was knocking down each door, trying to find the communications room. "Freeze!" a guard yelled, aiming a rocket launcher at him. Shift launched his fist at the guard's stomach, stunning him, and then kicked him into the wall, sending the guard to the next room, which had a sign on the busted door that read, "Communications."

The man at the computer jumped at the sight of the guard's unconscious body and turned away from the monitor with his hands up. Shift grabbed him by the neck and pinned him to the wall. "Did you or did you not block a call to the police department?" Shift said. The man had tears running down his face.

"Please, don't hurt me!" he cried.

"Answer me!" Shift said. He banged his fist next to the man's head, which dented the metal wall.

"Yes! Yes! A few days ago!" the man whimpered.

"Did someone send elimination orders?"

"Yes! A woman! In the Gilbert district," the man said. He coughed violently.

"Who ordered you to do this?"

"Oh God, no, please! He's going to kill me!"

"*You have three seconds!*" Shift yelled. "Three!" Shift tightened his grip on the man's neck. "Two!" Shift was ready to rip his throat out. "One!"

Finally, the man yelled, "His name is Cartwright!"

Chapter 18

Time stopped. The shots died. Shift looked out the door. All action was frozen. Shift released his grasp on the man's throat. Time resumed.

The man took deep breaths and coughed repeatedly. "Assistant Director Chris Cartwright. He gave no reason why she was to be terminated or why I had to block the call. I was just following orders," the man wheezed. After a few hard breaths, the man finally passed out.

Shift ran out the door. "Shock! Get us out of here!" he yelled. Shift's teammates ran to the docking station and huddled together. "Shock! Now!" Shift yelled. Shock was still dodging the bullets from the guards.

He was right by the entrance to the docking station, but the guards kept firing. He jumped and grabbed Shift's legs and teleported all five of them back to Base.

As soon as they landed, their armor melted away "This... was...as...far...as...I...could...get," Josh whispered.

"This is perfect," Aaron replied. Josh passed out from exhaustion. Aaron stood up and said to his teammates, "Can you guys look after him for a second? I need some time to thi—"

"Aaron!" said an adult male voice.

Aaron turned around. It was Cartwright. "You have some freaking explaining to do!" Aaron yelled. He approached Cartwright and stared into his eyes with a massive amount of hatred and confusion. "Why did you kill her?"

"I will explain. You all deserve it," Cartwright said. "You are all fascinating. I mean it! Extraordinaries have been living among us forever. Just recently, a collection of powerful men found out about your kind, they felt you were dangerous and thus founded the EPA. They worked to protect the world from your kind. That is when I came in. When we found out about you five, a secretive sub-organization within EPA was created and started work to make your kind into weapons. We discovered that your bodies are unlike ours. The human body is composed of organs and muscles and bones, but your bodies are made purely out of a shape-shifting material we called extraordinarium. We found you five and watched you. We supervised you in different atmospheres, like in school, under the park, and within these walls."

"Is everyone else at the school extraordinary?" asked Aaron.

"No," replied Cartwright. "If they were, we would train you all. The school is a decoy. A cover-up. Ever wondered why the other students were so lifeless? The EPA is rich in technology and funds. We used undercover agents and complicated animatronics to make the experiment pass as no more than a school. We used the school to lure you in. We thought that you would like to be a part of something normal for once. Except for Aaron, you had all lost your family, so we wanted to give you the chance to fit in with normal humans. After you started the training program, we realized that we did not need you, but only your extraordinarium. We took samples from you when you least suspected it. We tested it on humans, but they only turned into beastly creatures and then died. If we can figure out how to harness the extraordinarium, we can do unimaginable things for the world. If you are willing to give yourselves up for more experiments, we can mold the world into a much better place."

Cartwright held out his hand to Aaron. Aaron felt furious. Cartwright had been the villain all along; he had killed Aaron's mother, and now he wanted Aaron to join him? Aaron transformed and cut off Cartwright's left hand. Cartwright gasped in pain. Jade, Ethan, and Crystal transformed as well.

Cartwright slowly took out a syringe from his pocket. Inside was a gel that changed its shape and color constantly. "This is what your bodies are made of," said Cartwright. "This is just a few ounces of it. Imagine what we could do with all of it." He stabbed the syringe into his arm.

Cartwright stood motionless and then fell to the floor. He coughed up blood that was almost black. He started to transform. His whole body grew, but not his skin, which started to tear as Cartwright's massive new size pressed against it. Big, red blotches started to form on the rips in his skin, and a big, bleeding wart grew where his left hand used to be. Cartwright ripped his bloodstained, torn shirt off with his massive hand. He shrieked as he grew, until he was about twice his normal size. He looked like an enormous, zombie-like brute. He puffed, roared, and dashed toward Shift.

Cartwright swung his wart arm like a hammer at Shift, which sent him flying across the room and into the wall, causing it to crumble.

Eagle threw projectiles at Cartwright's eyes, and Cartwright raised his bleeding arm to block them. Cartwright ignored the projectiles and dashed toward Eagle. He jumped at the right moment, and Cartwright crashed into the wall. Eagle leaped onto his shoulders and started scratching the back of Cartwright's head with his talons. Cartwright recovered from the crash and punched Eagle, which sent him flying into the air.

Crystal ran up to Cartwright and bashed the bottom of his spine with a double fist. Cartwright turned around with a furious scowl. Crystal then roundhouse kicked Cartwright's jaw, which made him lose his balance and fall on his knees.

Moon Blade ran up to Cartwright, preparing her blade. She jumped and aimed it toward his shoulder, but Cartwright grabbed the blade with his good hand and threw the blade, along with its handler, to the side.

Shift had just recovered and spotted Cartwright getting back up on his feet. Shift made a grappling hook and threw it at Cartwright's leg. Shift pulled, tripping Cartwright, and sprinted toward him. He tackled Cartwright and made a buzz saw. Cartwright grabbed the saw, which barely cut through Cartwright's leathery skin. Cartwright tried to punch Shift with his wart-hammer, but Shift dodged it. Shift took his other fist and punched Cartwright several times. Cartwright twisted Shift's buzz-saw hand. Shift yelled in pain. Cartwright hammered Shift on the ground several times, and then he threw him across the room.

Nearby, Crystal pounded on the ground, sending several waves of pink crystals toward Cartwright. Two large crystals, similar to icicles, impaled Cartwright's legs, which made him stick in place. Crystal punched him several times, until Cartwright grabbed Crystal and flung her around as if she were a doll.

Moon Blade and Shift sprinted and tackled Cartwright.

Eagle came and slashed Cartwright's wrist, which freed Crystal.

Moon Blade hoisted her blade and stabbed it into Cartwright. Shift formed a hammer and, with all his might, smashed the other end of the blade, sending it through Cartwright's body. Cartwright roared in pain, since now he was completely impaled by the blade. Moon Blade grabbed her blade and pulled it out of Cartwright. Cartwright roared once again.

All the red blotches and warts started to pop, spilling blood everywhere. Cartwright coughed up blood and screamed until he shrunk back to his normal size. Shift,

Eagle, Moon Blade, and Crystal turned back to normal. Everyone went to see how Josh was doing except Aaron, who confronted Cartwright.

"I…didn't…want this," he said. "Gerald…ordered it."

"The superintendent?" asked Aaron.

"The director…of this whole thing," Cartwright replied. He coughed up some more blood. "He is…at the school. He…has a…hostage. I'm sorry. This…is the…only way to cleanse…the world." Cartwright gave one last huff of breath.

Cleanse? Aaron thought. He was still furious at Cartwright, but he accepted his last apology, whether he meant it or not. Aaron put his fingers on Cartwright's eyelids and closed them.

"Aaron! Josh is awake!" called Jade.

"What the hell happened?" asked Josh.

Aaron replied, "We got some filling in to do."

Suddenly, the turrets on the ceiling started to fire. Aaron and his friends quickly made their escape.

Chapter 19

It was an hour before dawn. Shift and his teammates were heading toward the school. Shift flew and carried Moon Blade, and Eagle carried Crystal with his talons, trying to be as gentle as ever. Shock had recovered and dashed through the air like an electric spark.

They got to the school and saw Gerald on the roof, next to a beaten Miss Birch, who was tied to a satellite. All five landed next to Gerald on the roof.

"I am impressed," he said. "I thought you would be dead by now. Fortunately for us, the EPA can pick you clean while you are still fresh. If you refuse, I will kill this woman."

"You can't fool us, Gerald!" Shift exclaimed. "We know this 'school' is just a fraud! She is not real!"

"Let's test that," said Gerald. He grabbed a silver pistol from his pocket and shot Birch in the leg. Her leg started to bleed.

"You sadistic maniac!" shouted Shift.

"We are not maniacs! We are intelligent beings, who happen to be smart enough to make you our puppets!" argued Gerald.

Shift said, "I could break you like a Popsicle stick! You will be put to—" A massive rumble filled the air. Strong winds started to blow. Shift looked up and saw a quad-copter, which was carrying a giant metal scorpion—the same one from in the underwater HQ! The helicopter landed the scorpion next to the school and flew away. Gerald concealed his awe from Shift.

"This"—he pointed at the scorpion—"is G—Genocide. We started this long before, but your training has helped us perfect this amazing piece of technology. Since you are unwilling to cooperate, this scorpion will make you, either by fear or force." Gerald slid down a ladder next to the school building. He approached Genocide and tapped a pattern on its eyes. The armor opened up, revealing a one-man capsule with many controls. Gerald climbed into it. The armor sealed, and Genocide started to boot up. It moved its legs first, then its claws, and finally its flexible stinger. "This world will change, whether anyone agrees or not!" shouted the speaker from Genocide. "It starts now!"

"Eagle, you get Birch out of here! The rest of us have no choice but to fight this thing!" said Shift.

Eagle grabbed Birch and flew away. He quickly placed her on top of a building nearby. "Stay here," he said. Birch nodded. Eagle flew back to his teammates to listen to the plan.

"This is just like training, guys. We need one person at the torso, two at the legs, one at the claws, and one at the stinger. Be cautious. We don't know about any innovations this thing has," said Shift.

Genocide took its right claw and smashed the wall of the first floor of the building. Eagle jumped at the stinger. Moon Blade slid down the crumbling rocks and attacked the claws. Crystal pounded the right legs of Genocide, and Shock zoomed around the left legs, punching and kicking at each leg.

Shift jumped and dived at Genocide's torso. He landed and scouted for a weak spot. Three guns ejected out of the armor and shot at Shift, who had already prepared a shield with spikes on the surface. He charged at the guns and the spikes destroyed them. All that remained were the three posts that the guns stood on. A whirring sound began to shake the area. The posts magnetically attracted several metal objects which floated in the air, broke into little metal pieces, and started connecting to build three new turrets that started shooting immediately at Shift, who was unprepared. He was knocked back, hit a brick wall, and crashed into a classroom. *I forgot about that rebuilding ability*, he said to himself.

Eagle flew around the stinger, dodging and slicing it with his wings. The stinger struck him and went through his armored body and into his skin. The inside of the needle stinger started to widen, forming a tube to suction out some of the extraordinarium from Eagle's body. He yelled and pulled the stinger out of his body.

Crystal was attacking the legs, which tried to pin her to the ground. After a few attempts, the middle leg was successful and started to absorb her extraordinarium. Shock spotted her and tried to help. He slid under the torso of Genocide, but metal rings ejected from the scorpion's legs and tied on to Shock's arms. His wrists were locked together, and he could not break free. Crystal was still being absorbed. Shock could only think of one thing to do, but it was far too risky; however, it was his only hope. He ran up to the leg and started kicking it. "Hey, you rusty pipe! I'm here!" The leg retracted its stinger from Crystal and shot it at Shock. As he was being drained, Shock surged himself with electricity. The amount of power he was producing was painful and overwhelming, but it blew up the leg. The place where the leg used to be immediately magnetized the closest piece of metal it could find—the metal rings on Shock's wrists—in attempt to rebuild its leg. Before it could rebuild, Crystal smashed the skinny leg, making microscopic bits fly.

Meanwhile, Moon Blade was dodging and flipping over the claws, occasionally slicing Genocide's surface. The left claw snapped at her, but she took one of her blades and placed it in between the right claw. The claw tried to snap shut, but ended up breaking itself. Moon Blade got her blade back and tried stabbing Genocide's eyes. She made no impact and fell to the floor. The left claw picked her up and injected needles into her skin. At the last moment, Shift threw a spiked boomerang at the claw, which released its grip. Moon

Blade fell to the ground and Shift escaped with her. He flew to the top of the school and looked at his teammates. He saw them fighting their best, but they were being over powered by Genocide and drained of their extraordinarium. He heard their gasps and cries for help and time seemed to freeze. Shift had an idea. "Stay here," he told Moon Blade.

Shift flew as fast as he could to retrieve his teammates from Genocide's clutches. He carried them all, who were too worn out to move. He flew back to the roof and gave directions. A few protests were given, but everyone acquiesced. They turned back to normal, stood up, and held hands. "Hey, Gerald!" Aaron shouted. Genocide turned around and inspected the five.

"Have you given up already?" he asked, followed by a sinister chuckle.

"You said you wanted us, so come get us, you bastard!" Genocide prepared itself by ejecting five needles out of its stinger. Aaron looked at his teammates. His partners. His friends. They were all beaten and weary and were covered in leaking extraordinarium. He squeezed his left hand, which held Ethan's hand, and squeezed his right, which held Jade's. They all stood tall.

The stinger struck at lightning speed, like a snake going for its prey. Without their armored alter egos, the five felt even more pain and were drained even faster. They fell to their knees, but still held hands. *"Tank overload! Tank overload!"* a robotic voice coming from Genocide warned. Genocide's abdomen started to spurt out gas, liquid, and chunks of solid all at once. Aaron guessed that this was extraordinarium. Genocide's armor and limbs started to come off.

After the whole armor shell fell off, the entire scorpion blew up. Fire clouds flew everywhere, knocking Aaron and his friends off the roof and onto the ground. Aaron retrieved his awareness and quickly gathered his friends to safety.

After a few minutes, most of the fire and smoke had cleared, except the control capsule. Gerald was crawling on the ground. His skin was scorched. Aaron ran in and grabbed him out of the fire. They brought him to the entrance of the building and lay him on the door. They noticed that both his legs had been blown off clean. "This will not—" said Gerald. He gave a dry, violent cough before continuing. "This will not stop anything. The world…will still hate you. The EPA…will remain. There is no one…left who loves you." Aaron was about to punch Gerald, thinking that his body was brittle enough now to smash. He had thought about how Gerald gave the order to kill his mom and how killing Gerald would be a waste of the little energy Aaron had left.

He stood up and hugged his friends, leaving Gerald where he was. "We did it," he whispered. He felt more joy than he had ever felt before. He and his friends helped each other up started to limp away. Ethan had his arms over Aaron's and Jade's necks, since he did not have his prosthetic feet. Everything was peaceful—until a bullet whirred through the air and into Gerald's head.

No one had any energy to fight, and they could not protect themselves. Aaron heard more bullets whir through the air and into his teammates. They hit Josh first, then Crystal, then Ethan, and then Jade, who all fell to the ground. He gazed over the bodies and was ready to scream until he spied something on Josh's neck. It wasn't a bullet. It was a tranquilizer dart. Before he had time to think, he fell to the ground and everything went dark.

Chapter 20

What's happening? Where am I? Where are the others? Aaron found himself surrounded by darkness except for a single glowing beam of light that started to grow. It showed a vision of a man with a woman—his wife, maybe? The woman held a small baby, weeks old at the most. A small girl, probably the age of two, tugged on the man's pants. "Daddy?" asked the girl. "When will you be back?"

The man knelt down and put his hands on her shoulders. "Who said I was leaving?" he said in the calmest tone. "I will always be with you, Mommy, and your brother." He kissed the girl's cheek, and then took the baby from the woman. He held the baby for a bit and then gave the baby back to the woman. "Make sure neither of them find out. I love you, Dee," the man whispered to the woman.

Aaron wanted to speak with them, but he couldn't. He heard a crash and loud footsteps. The man turned. Rushing up a set of stairs were two soldiers, who had navy-blue helmets and uniforms. They started to beat the man until he was unconscious. The girl ran to the man, but Dee held her back. The guards grabbed his lifeless body and moved downstairs. Aaron wanted to run and get the soldiers, but instead, his whole vision darkened again. Aaron thrust his body upward and fell off a bed.

He looked around. He was in a hospital room. The whole room was white, gray, and black. He noticed that he was wearing clean clothes the same color scheme as the room. He also had stitches in several places on his body. Aaron had fallen off a white bed and onto a gray-tiled floor. He stood up and looked around more. The room had no windows. The entire room was the size of a large closet. He found a door and opened it. Before he left, he peeked through the crack of the door. The hallways were the same color scheme. He tiptoed out of the room and walked to the end of the hallway.

He looked around the corner and saw his friends, sitting down and eating. They were dressed in the same white, black, and gray clothes. Ethan even had two new silver prosthetic feet that resembled tennis shoes. Aaron was about to greet his friends when a woman's voice behind him said, "I am aware that your healing should be rapid, but we put those stitches on just in case." Aaron almost jumped. It was Miss Birch.

She had bandages on her leg, and she, too, was in a uniform, except her uniform was more elegant and less prison like. The strange thing was that she was walking with only a limp, as if she had only sprained her ankle. "Calm down.

You're safe," she said. She guided him to the next room, where his friends greeted him. "I know you may have some trust issues, but after I explain, I guarantee you'll understand."

Aaron sat down with his friends, who welcomed him, and listened. "You are all lucky that you had enough extraordinarium left in your bodies to regenerate itself and keep you alive."

"You're not a doctor, are you? Probably not a teacher either," asked Aaron sarcastically.

"Negative, but I will reveal my true identity. I am not Miss Birch. I was undercover, watching you all. I am Director Sava. I am the head of Cooperation with Extraordinary Beings Agency, or CEBA. We will not study you, betray you, or do any of the horrible things the EPA did. Let's just say that we fight EPA's efforts. The assassination of Gerald helped us apprehend several EPA-held bases and property, including your training base, your underground housing base, the school, and the Andersons' house."

"You mean we can live there after all?" asked Aaron excitedly.

"Eventually. You will need to stay here for a few weeks under our supervision. We will be rebuilding the school, filling it with real kids. That will take at least a year. Until then, you will be receiving real education and better training here," she answered.

It took a while for them to process this. Aaron thought of a question that was really just a tangent, but it had to be answered. "Are we the only extraordinaries left, or did Gerald and Cartwright lie to us?"

"They lied," Sava answered. Aaron exchanged confused looks with his friends.

"There are more extraordinaries out there. Some hate the humans back, thus forming a worldwide organization led by a powerful extraordinary, much more powerful than you all. His powers…are identical to Aaron's, but much stronger and much better controlled, so he can do more than just shape-shift. He can shape-shift his powers."

Everyone looked at Aaron. "Wonder if we're related," Aaron joked.

"He leads terrorist attacks across human cities along with others who hate extraordinaries. There are several sub-organizations that lead other groups of extraordinaries with a variety of abilities. You will be trained to negotiate and, if necessary, fight against them," said Director Sava. Everyone got excited for several reasons, like not being alone and being able to interact with the real world.

"Wait. Where are we exactly? You never told us," asked Crystal.

"Right!" Director Sava said. "We are aboard the SS *Extraordinary*." She walked to the side of the room and said her name. A laser from above scanned her. The wall slowly disappeared, leaving a door-sized window. They all got up and looked outside. They saw the entire horizon of the city. The light of the sun stood out from the mostly dark horizon and slowly grew, just like the future of Aaron and his friends.

ACKNOWLEDGMENTS

Countless people, including my family and my editor, and circumstances contributed to this project. I am so grateful for everyone's help, whether they gave bits and pieces of inspiration or massive chunks. Without these contributions, my biggest dream would have never come true.